“Here is The Bike Club.
I’ll go in,” said Pete the Pig.

"Do you like to ride?"

the pigs asked.

The nine pigs made Pete smile.

"We'll ride a mile!"

"These are two bike rules,"
said the pigs.
"Don't speed. Don't pass."
"I won't," said Pete the Pig.

The bike path was long
and wide.
Pete the Pig had a fine ride.

"Do you see Pete the Pig?"
asked one pig.
"He's doing fine.
He'll like The Bike Club."

“Drinks! One dime! Line up!”
said a pig.
“Get your drinks here.”

Then it was time to go.
"We'll ride back to The Bike
Club," said the pigs.

"Here," said the pigs.

"Take these."

"Thanks, friends!"

said Pete the Pig.

"I had the best time.

I'll be back!"

The End